YOUR KNOWLEDGE HAS VALUE

- We will publish your bachelor's and
  master's thesis, essays and papers

- Your own eBook and book -
  sold worldwide in all relevant shops

- Earn money with each sale

Upload your text at www.GRIN.com
and publish for free

Bibliographic information published by the German National Library:

The German National Library lists this publication in the National Bibliography; detailed bibliographic data are available on the Internet at http://dnb.dnb.de .

Imprint:

Copyright © 2018 GRIN Verlag
Print and binding: Books on Demand GmbH, Norderstedt Germany
ISBN: 9783668795488

This book at GRIN:

https://www.grin.com/document/437683

Saad Khalil, Mohamad K. Aroua

Effects on Surface Area. Intake Capacity and Regeneration of Monoethanolamine

Palm-Shell Activated Carbon Prepared for CO2 Adsorption

GRIN Verlag

Effects on surface area, intake capacity and regeneration of monoethanolamine and 2-amino-2-methyl-1-propanol impregnated palm-shell activated carbon prepared for CO_2 adsorption

SAAD HASHIM KHALIL

Chemical Engineering Department. University of Malaya, Kuala Lumpur, Malaysia

Abstract

Granular palm shell activated carbon (AC) was impregnated separately with monoethanolamine (MEA) and 2-amino-2-methyl-1-propanol (AMP) to improve its natural capacity and selectivity for carbon dioxide (CO_2) adsorption. The total surface area, micropore volume, as well as the heterogeneity of the impregnated AC particles was considerably reduced due to impregnation. CO_2 intake of impregnated 500 μm AC particles improved significantly and adsorptive capacity of 500 μm MEA-impregnated AC particles improved by 172 % and 44 % comparing to non-impregnated and AMP-impregnated AC particles respectively. Solid state amine stoichiometric results indicated that adsorption capacity of unhindered amine (MEA) is higher than that of hindered amine (AMP) by 50 % contrary to liquid amines standard stoichiometry. Exhausted AMP-impregnated beds were regenerated by sweeping at room temperature with stream of pure nitrogen (N_2) flowing at 60 ml/min for 4 hours. Heating up to 75 °C was required to regenerate exhausted MEA-impregnated beds. Increasing feed gas flow rate has adverse effect on breakthrough time more than increasing bed operating temperature. Breakthrough time was utilized to evaluate the performance of the different adsorption beds.

Keywords: Activated carbon; CO_2; Adsorption; Impregnation; MEA; AMP.

1. Introduction

The topics of climate change and global warming have been discussed and investigated thoroughly. While both of them are naturally occurring phenomena, climate change is the average long-term changes in climate. Global warming, which is one of the causes of climate change is the increase in the temperature of the lower atmosphere due to the effects of greenhouse gases. Global warming occurs naturally in cycles throughout periods of time due to changes in the profile of the Earth path around the sun known as Milankovitch cycles, but the current episode of global warming is believed mostly to be man-made [1] because of the unprecedented increase in the concentration of greenhouse gases above the earth surface particularly due to the anthropogenic emissions of the industrial revolution, which had added substantial quantities of greenhouse gases (GHGs) to the atmosphere. GHGs like water vapor, carbon dioxide, methane, nitrous oxide, and other minor atmospheric gases absorb some of the outgoing infrared radiation (heat) reflected by the earth surface. The retained heat would accumulate and raise the lower atmosphere temperature [2]. The most important anthropogenic GHG is CO_2, which is virtually considered responsible for causing the greenhouse impact [3]. Harmful effects of changes in precipitation and temperature is partly believed to be due to the GHGs mainly as a result of fossil fuel consumption such as coal, oil and natural gas in daily life activities, which is providing more than 85% of global energy needs [4], but at the same time extensive relying on fossil fuels increased the average atmospheric temperature by 0.8 °C in the last 150 years [5]. To reduce CO_2 concentration in the atmosphere, various technical options, such as adsorption [6], absorption [7], membrane separation [8] and cryogenic [9] have been proposed and investigated for the capture and sequestration of CO_2. Because of low energy requirement, minimal effects on environment, and ease of applicability, adsorption is considered as one of the most capable techniques in the commercial and industrial applications [10]. Solid sorbents, like AC is one of the most auspicious options for post-combustion CO_2 adsorption [11]. Porous AC is produced from a wide range of carbon-rich raw materials and biomass resources, which is considered renewable [12]. AC texture is adaptable as the surface functional groups trait and density as well as the pore size distribution can be adjusted to suit the application [13]. Microporous AC is superior adsorbent due to its high internal surface area, where the most of the adsorption occurs, narrow pores, where the adsorption capacity is significantly improved, and the presence of chemically active sites, which are vital for the adsorption of specific groups [14]. It is well known that the ability of ACs to adsorb acidic gases will increase when there are nitrogen functional groups on its surface [15]. Modification of AC's surface chemistry for gas and liquid adsorption by impregnation via slurry method are widely used, albeit surface area reduction and pore blockage are common occurring obstacles [16]. CO_2 capture capacity by modified ACs was found to be improved due to the introduction of alkaline nitrogen groups on their surface [17]. Calcium acetate solution blocked the micropores of AC particles when they were impregnated onto AC samples reducing their surface area but in the same time enhancing CO_2 adsorption by the basic sites formed on the AC surface [18]. Adsorption of CO_2 on impregnated palm shell AC particles is limited in literatures [19] and the topic is even more limited on the dynamic CO_2 adsorption onto palm shell AC particles. Industrial alkanolamine solutions used to absorb CO_2 are, monochtanolamine (MEA), diethanolamine (DEA), N-mehyldiethanolamine (MDEA), piperazine (PZ) and 2-Amino-2-methyl-1-propanol (AMP). MEA is considered the most effective among the other mentioned alkanolamines [20] because of its high reaction rate. Primary amine MEA is considered frugal [21] and has a higher absorption rate than other secondary, tertiary and sterically hindered amines in the order: MEA > AMP > DEA ≫MDEA [20], but there are some shortcomings in using MEA and AMP as absorbents due to their corrosive behavior, moderate biodegrability and toxicity [21]. In the meantime CO_2 capture technologies available for carbon capture and sequestration (CCS) are pre-combustion, post-combustion and oxyfuel or oxy-combustion [22]. For the first and second systems as the name implies, CO_2 is removed before or after a fuel is burned respectively, while the third system is using pure oxygen instead of air for combustion. Post-combustion CO_2 capture is offering significant advantages over the other two systems due to its readiness to be used in many fossil fuel-fired power plants scattering around the world without the need to modify the combustion process and the other technologies involved, which is usually expensive and extensive [23].

In this work, fixed bed packed with granular AC particles, which are cost-effective adsorbent and its raw materials abundantly available in Southeast Asia particularly Malaysia [24], impregnated separately with primary (MEA) and sterically hindered (AMP) amines, was employed in a dynamic real time experiments to adsorb CO_2 from gas mixture. Water was used as ecologically friendly medium to facilitate the impregnation of the amines. Dynamic breakthrough time plays major rule in this research as a tool for examining and evaluating the beds adsorption quality.

2. Materials and Methodology

2.1. Materials

2.1.1.

Analytical grade monoethanolamine and 2-amino-2-methyl-1-propanol were used in this study, as shown in Table 1.

Table 1 Chemical formulas and molecular weight of MEA and AMP

Amine Based Chemical	Chemical Formula	Molecular Weight, g/mol
Monoethanolamine (MEA)	C_2H_7NO	61
2-amino-2-methyl-1-propanol (AMP)	$C_4H_{11}NO$	89

2.1.2. Palm shell AC

Commercial granular palm-shell AC (particle size, 710-500 μm), which was produced by physical activation with steam as the activating agent was purchased from Bravo Green SDN BHD

(Sarawak, Malaysia). Total BET surface area was found to be 838 cm^2/g. Energy Dispersive X-ray (EDX) elements microanalysis showed that the surface of the AC particles is comprised of 88.89 % carbon, 7.75 % oxygen, 2.81 % silica, 0.14 % aluminum and 0.41 % nitrogen by weight.

2.1.3. Gases

Gases used in this study were:

1. Mixture of 15% CO_2 and 85% N_2.

2. Pure N_2.

2.2. Methodology

2.2.1. AC samples characterization

AC particles were crushed and sieved. Sieves of the sizes of 850, 710 and 500 μm were employed to characterize the AC particles. Two samples were obtained, namely: 710 μm (particles passing 850 and stopping on 710 μm) and 500 μm (particles passing 710 and stopping on 500 μm).

2.2.2 MEA and AMP selection

MEA and AMP were selected among other amines because, MEA is considered as one of the most mature amine absorbents for CO_2 absorption [25], and AMP due to its less energy requirement for regeneration than other amines [26].

2.2.3. Impregnation of AC samples

Impregnation was carried out by stirring 2 g of each MEA and AMP with 5 g of AC for 1 hour, 10 g of deionized water was added as an environmentally friendly medium to facilitate the impregnation process. The final slurry was dried at 70 °C, under vacuum pressure (- 0.1 bar) for 6 hours. Loading efficiency was found to be 91% for both samples.

2.2.4. Adsorption of CO_2

CO_2 molecules were adsorbed onto a combination of solid state MEA and AMP impregnated separately on mostly micropore palm shell AC particles. Solid state amines employment was essential to minimize the harmful effects of liquid amines on environment and to curb their corrosive impacts and their degradation and the high energy required during regeneration [27].

2.2.5. Evaluation method of the adsorption beds performance

Breakthrough time was used to evaluate the adsorption beds performance in real time experiments. Breakthrough time is an expression in time unit standing for the adsorption bed capacity and quality. It can be defined as the elapsed time for adsorption bed to be fully or partially saturated with adsorbate molecules as in this study when CO_2 molecules break through the adsorption bed, which was monitored by Guardian Plus CO_2 monitor. Data Acquisition Logger was connected to the CO_2 monitor to measure the breakthrough time in minutes.

2.2.6. Experimental setup

Experimantal setup is shown in Figure 1. Details of experimental setup's apparatuses and their specifications can be found in [28].

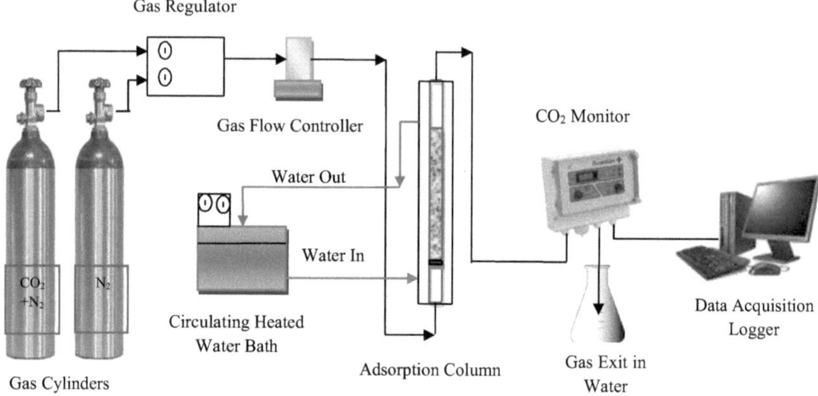

Figure 1 Experimental Setup

3. Results and Discussion

3.1. Breakthrough time results

Results in Figure 2 are showing the differences in breakthrough time for three adsorption beds at room temperature. Adsorption of CO_2 molecules is considered concluded when the display of the CO_2 monitor, which is connected to the outlet of the adsorption column, is showing that CO_2 molecules start to exit the bed. These results were obtained using 5 g of each adsorbent, at room temperature and 10 ml/min feed gas flow rate.

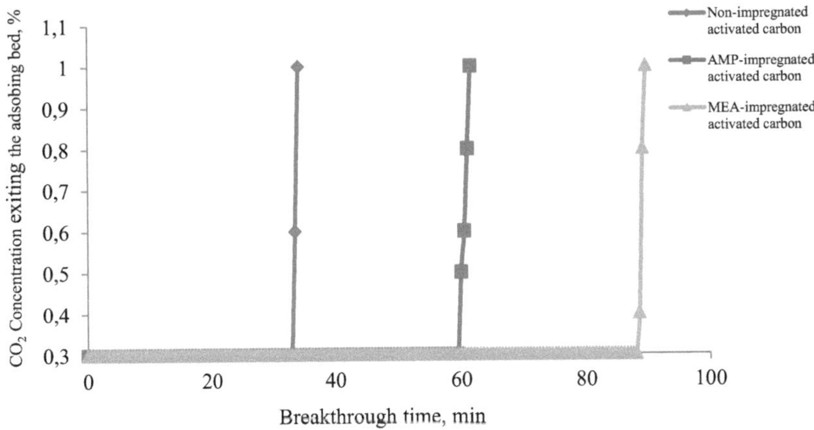

Figure 2 Breakthrough time at room temperature for 500 µm, non-impregnated, AMP impregnated and MEA-impregnated AC beds

Breakthrough time values in Figure 2, increased from 34 min for non-impregnated AC bed to 62 min and to 90 min for AMP-impregnated for MEA-impregnated AC bed respectively. The increase in breakthrough time for MEA and AMP-impregnated beds compared with non-impregnated AC bed was due to that amine molecules have formed many active sites for CO_2 adsorption, where chemisorption was dominant over physisorption leading to adsorb selectively more CO_2 molecules from the feed gas stream. Steric hindrance influence is responsible for breakthrough time reduction of AMP-impregnated AC beds in comparison with MEA-impregnated AC beds.

3.2 Steric hindrance effect

Breakthrough time results are showing that MEA-impregnated AC beds have extended breakthrough time comparing to AMP-impregnated AC beds because of the steric hindrance effect on large amine molecules [29], like AMP, which would hinder CO_2-N_2 reaction causing more CO_2 molecules to leave the adsorption bed, hence reducing breakthrough time as shown in Table 2.

Table 2 Adsorbing beds and their breakthrough time

Adsorbing bed	Breakthrough time (min)
MEM-impregnated AC	90
AMP-impregnated AC	62
Non-impregnated AC	34

3.3. Dynamic adsorption beds capacity improvement

The beds dynamic adsorption capacity, mg/g. (mg of CO_2 adsorbed / g of adsorption bed), was calculated using equation 1, for the period identified from the beginning of the adsorption experiment when CO_2 molecules first flow through the adsorption bed until their breakthrough out of the bed ending the adsorption experiment.

$$Capacity = Q\rho Ct/W \qquad\qquad 1$$

Where:

Q = Volumetric feed gas flow rate (ml/min)

ρ = CO_2 density at 25 °C (mg/ml)

C = CO_2 (%) in the feed gas

t = Breakthrough time (min)

W = Weight of amine-impregnated bed (g)

Results in Table 3, are showing the improvement of MEA-impregnated and AMP-impregnated bed comparing to non-impregnated bed.

Table 3 Adsorption bed capacity improvement of MEA-impregnated and AMP-impregnated bed compared to non-impregnated bed

Adsorption bed	Adsorption capacity, mg/g	Adsorption capacity improvement, %
Non-impregnated (500 µm) AC	18	172
AC (500 µm) impregnated with MEA	49	
Non-impregnated (500 µm) AC	18	89
AC (500 µm) impregnated with AMP	34	

3.4 Unhindered MEA and hindered AMP stoichiometry

Experimental stoichiometric results showed that 0.6 mole of CO_2 reacted with 1 g AMP, while 0.9 mole of CO_2 reacted with 1 g MEA, an 50% increases in MEA capacity for CO_2 adsorption. These results are inconsistent with the liquid amine theoretical stoichiometric, which states that at room temperature and pressure the maximum amount of CO_2 reacting with unhindered amine (MEA) is 0.5 moles for each 1 mole of amine, and a maximum of 1 mole CO_2 reacting with each 1 mole of sterically hindered amine (AMP) [30], which was confirmed by many researchers [31-33]. This inconsistency is occurring due to the absence of water. The outcome of CO_2-AMP (hindered amine) reaction is unstable carbamate ions, which would be hydrolyzed to more stable bicarbonate ions and free amine [34]. The formation of free amine would enhance the AMP absorption capacity by facilitating more CO_2 absorption, but in the absence of water no free amine would be formed hence reducing the adsorption capacity of AMP comparing to unhindered MEA, which when react with CO_2 would form stable carbamate [30].

3.5. Adsorption isotherm of impregnated and non-impregnated 500 μm AC beds

Specific surface area of impregnated and non-impregnated 500 μm AC beds was calculated by the adsorption isotherm method using standard Brunauer–Emmett–Teller (BET) equation; the micropore area and volume were obtained using the *t*-plot method of Lippens and de Boer. Pores size and volume were measured using Barrett-Joyner-Halenda (BJH) pore analysis method.

Three adsorption isotherms of 500 μm, non-impregnated, MEA-impregnated, and AMP-impregnated AC samples were plotted. The adsorption isotherms and BET surface area were calculated using Micromeritics Accelerated Surface Area and Porosimetry Analyzer (ASAP 2020), with N_2 adsorption experiments at liquid N_2 temperature of -196 °C. The samples were degassed at 130 °C for 1 hour in a vacuum environment before the N_2 adsorption measurements. BET surface area of AC beds impregnated with MEA and AMP was found to be (65 m^2/g) for both beds, comparing to (838 m^2/g) for non-impregnated AC bed. The significant reduction in the surface area of the impregnated AC beds was due to MEA and AMP molecules blocked the mainly micro-size pores AC particles. The blockage of large number of micropores in the AC particles by the impregnating amine molecules had left high number of unblocked mesopores and the ratio of mesopore/total surface area in the MEA, AMP-impregnated AC beds were almost twofold of that of non-impregnated AC beds comparing to non-impregnated beds due to impregnation. BET surface area of the three adsorbents is shown in Table 4.

Table 4 Total micropore and mesopore surface area for different beds

Bed	Total surface area, m^2/g	Micropore surface area, m^2/g	Mesopore surface area, m^2/g	Mesopore/total surface area, %
Non-impregnated	838	675	163	24
MEA-impregnated	65	36	29	45
AMP-impregnated	65	38	27	42

Standard classification of adsorption isotherms is showing in Figure 3. Class I is for adsorbents with a predominantly micropore adsorbents and class II is mostly for mesopore adsorbents, where the hysteresis loops appear as a common feature associated with the presence of mesoporosity due to the condensation of adsorbate molecules inside the mesopores [35].

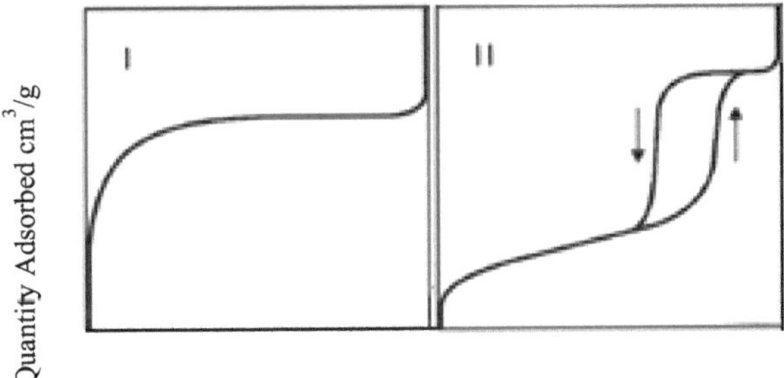

Figure 3 Classification of adsorption isotherm [35]

Adsorption isotherm of non-impregnated 500 µm AC sample is showing in Figure 4, which is corresponding to type I isotherm. In micropores, the potential of both sides of the pore walls overlaps. This overlapping would enhance the adsorption capability of the adsorbent, where adsorption starts almost instantly when N_2 molecules contact the AC particles [36].

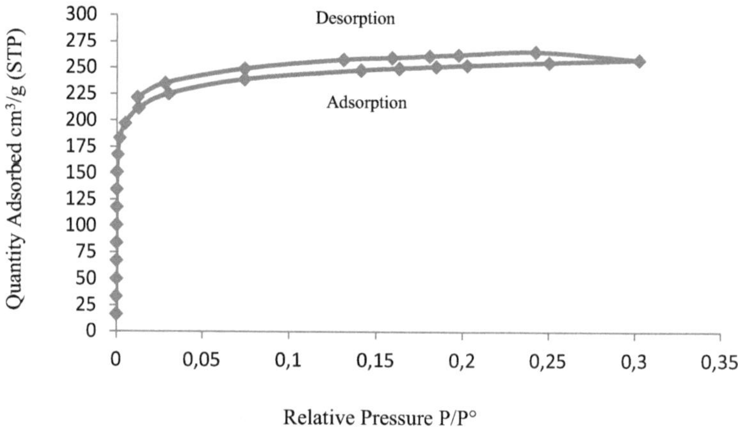

Figure 4 N_2 adsorbed onto 500µm, non-impregnated AC particles at 77 K

In Figure 5, the adsorption isotherm of 500 µm AC sample impregnated with MEA is corresponding to type II isotherm

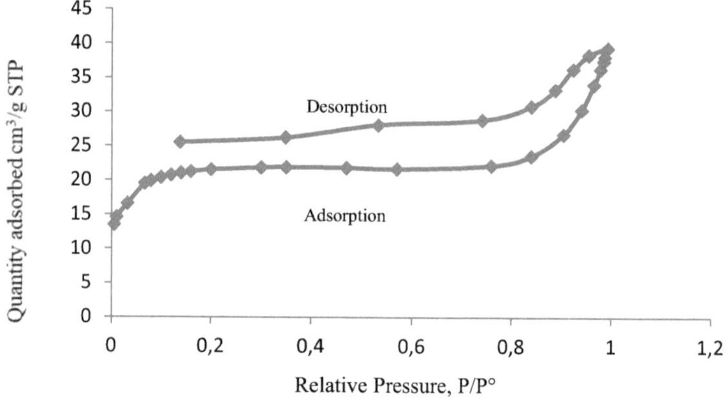

Figure 5 N_2 adsorbed onto 500 m, AC particles impregnated with MEA at 77 K

Figure 6 is showing the amount of N_2 molecules, adsorbed onto the 500 µm AC particles impregnated with AMP, which is following the same trend as in Figure 5.

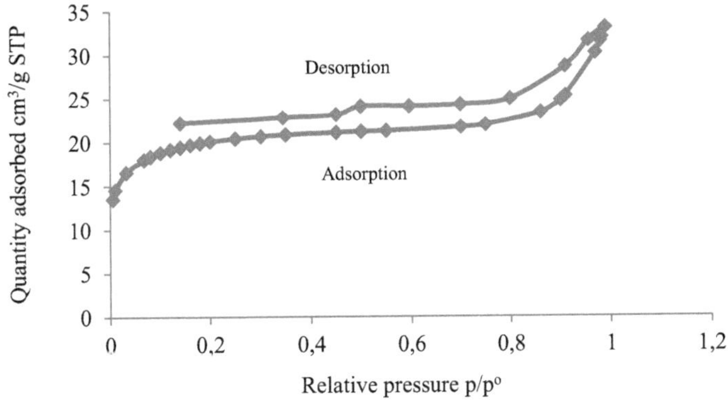

Figure 6 N_2 adsorbed onto 500 µm, AC particles impregnated with AMP at 77 K

Due to impregnation with amines, the total micropore volume of the AC particles decreased by 94 % for both MEA-impregnated and AMP-impregnated AC beds comparing to non-impregnated beds as shown in Table 5.

Table 5 Micropore volume for different adsorption samples

Adsorption bed	Microporepore volume, cm³/g
Non-impregnated AC	0.32
MEA-impregnated AC	0.02
AMP-impregnated AC	0.02

3.6. Linear regression of Dubinin-Astakhov (D-A) equation

A linear regression was performed on (D-A) equation for non-impregnated, MEA-impregnated and AMP-impregnated 500 µm AC samples from adsorption isotherms data obtained by ASAP 2020.

Equation 2 is D-A equation [37]:

$$w = w_o e^{-(A/E)^n} \hspace{4cm} 2$$

Where:

W = Amount of gas adsorbed (cm^3/g).

W_o = Micropores volume (cm^3/g).

A = Adsorption potential (J/mol).

E = Characteristic energy of adsorption (J/mol).

n = Structural heterogeneity parameter.

The adsorption potential defines as in equation 3:

$$A = RT\ln\frac{p_o}{p} \hspace{4cm} 3$$

Where:

P_o = Saturation vapor pressure of the adsorbate.

P = Pressure in the gas phase.

Substituting equation 3 in equation 2:

$$w = w_o e^{-[(RT/E)^n (\ln{^{p_o}/_p})^n]} \hspace{3cm} 4$$

The linearized form of equation 4 is as in equation 5:

$$\ln w = \ln w_o - (RT/E)^n \; [\ln(p_o/p)]^n \hspace{2.5cm} 5$$

Equation 5 is another form of equation D-A and it describes adsorption on both homogeneous and heterogeneous microporous adsorbents. The degree of heterogeneity of an adsorbent can be estimated by the value of exponent (n). The value of exponent n in microporous ACs is ranging from 1.5 to 3, as the adsorbent becomes less heterogeneous [38]. Equation 5 is plotted in Figures 7, 8 and 9 for non-impregnated, MEA-impregnated, and AMP-impregnated AC samples respectively. Coefficient of determination (R^2) was employed to select the best value of exponent n, as shown in Table 6. The structure of AC is very complicated and its surface is highly heterogeneous [39]. That was confirmed by the value of exponent n (1.5) for non-impregnated AC sample, which shows that the sample is more heterogeneous than both MEA-impregnated and AMP-impregnated samples with their exponent n value equals to 2, where in this particular case equation 3 would be called Dubinin-Radushkevich equation (D-R). The results are shown in Figures 7-9 and Table 6. To explain these results, Table 4 earlier was showing that micropore volume of non-impregnated AC beds is greater than that of MEA and AMP-impregnated AC beds, which can be interpreted as that the micropores in non-impregnated sample are more diverse and there are more variations in their shapes and sizes than MEA and AMP-impregnated beds, where many of their micropores are blocked by MEA and AMP molecules, reducing their pores diversity.

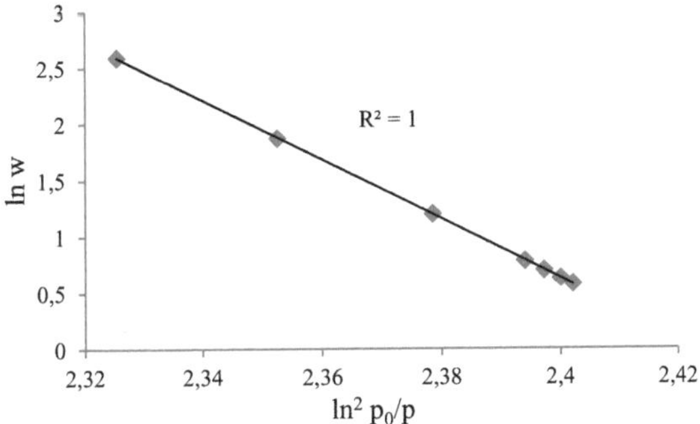

Figure 7 Non-impregnated, 500 μm AC sample

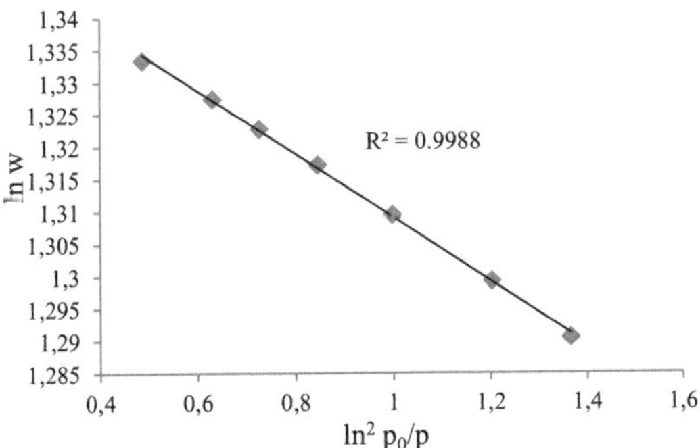

Figure 8 MEA-impregnated, 500 μm AC sample

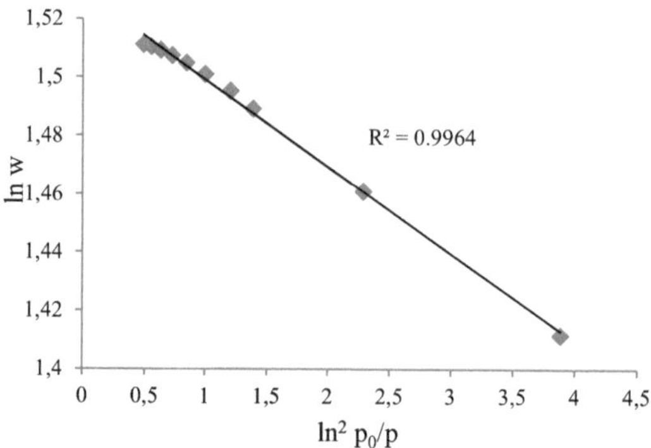

Figure 9 AMP-impregnated, 500 µm AC sample

Summary of the results in Figures, 7, 8 and 9 are presented in Table 6:

Table 6 Results of exponent n and coefficient of determination (R^2) for different samples

Sample	Linear equation	n	R^2
Non-impregnated	y = -0.025x+5.5814	1.5	1
MEA-impregnated	y = -0.0214x+3.1277	2	0.9988
AMP-impregnated	y = -0.0129x+3.5206	2	0.9964

3.7. Thermal characterization of the AC beds

Adsorption is mostly exothermic process [40] since new bonds would be formed between the adsorbate and the adsorbent. Thermal probe connected to digital display was inserted inside the AC bed to record the temperature of adsorption. Experiments were performed on 5 g of each of non-impregnated, MEA and AMP-impregnated AC beds; feed gas flow rate was 10 ml/min at room temperature. During the adsorption experiments, non-impregnated AC bed showed slight increase in temperature due to the weak physisorption van der Waals attraction forces that formed between CO_2 molecules and the non-impregnated AC particles.

There was further increase in temperature for MEA and AMP-impregnated AC beds due to the establishment of chemical bond between CO_2 molecules as weak acid anhydrite [41] and the MEA and AMP molecules as a weak base. Results are shown in Table 7.

Table 7 Increasing in bed temperature during adsorption

Type bed	Increase in bed temperature, °C
Non-impregnated AC	0.5-1.5
MEA -impregnated AC	5
AMP -impregnated AC	7

3.8. Effect of regeneration on the performance of 500 µm AMP-impregnated AC beds

Two cycles of regeneration were performed by sweeping the exhausted bed upward with 60 ml/min pure N_2 at room temperature, for 4 hours. Regeneration in Table 8 was carried out at room temperature, results can be improved if heat was to be used. Heating would provide enough energy required to sever the N_2-CO_2 chemiacl bond. Results in Table 8, suggesting that AC beds impregnated with AMP were regenerated with reasonable reduction in breakthroug time especially for 500 µm beds.

Table 8 Reduction of breakthrough time of 500 µm AMP-impregnated AC before and after
regeneration

Breakthrough time AMP-impregnated AC, min	Breakthrough time AMP-impregnated AC after regeneration, min	Reduction in breakthroug h time, %
62	58	6
60	55	8

3.9. Effect of regeneration with high temperature on the performance of MEA- impregnated AC beds

Regeneration of exhausted MEA-impregnated AC beds would be processed by sweeping the exhausted beds with pure N_2 gas flowing at 60 ml/min for 4 h. Water bath was used to heat the exhausted bed in situ up to 75 °C by circulating hot water through the jacket surrounding the exhausted bed at the same time as N_2 is sweeping the bed. Results in Tables 9 and 10 are showing that regeneration with heating up the exhausted beds is more efficient than regeneration at room temperature. This is due to that MEA-CO_2 bond is more established than APM-CO_2 bond [30].

Table 9 Breakthrough time reduction of MEA-impregnated AC beds at room temperature before
and after regeneration

Particle's size, µm	Breakthrough time, min	Breakthrough time after regeneration, min	Breakthrough time reduction , %
710	52	26	50
500	85	40	53

Table 10 Breakthrough time reduction in of MEA-impregnated AC beds, heated to at 75 °C before and after regeneration

Particle's size	Breakthrough time, min	Breakthrough time after regeneration, min	Breakthrough time reduction , %
710	54	47	13
500	87	72	17

3.10. Effect of bed operating temperature on bed's breakthrough time

Figure 10 is showing the effect of increasing bed operating temperature on breakthrough time of three different adsorption beds. Due to the exothermic nature of adsorption, increasing operating temperature would reduce breakthrough time. This effect would be limited on non-impregnated AC beds as these beds trapping CO_2 molecules physically by the van der Walls forces. For MEA and AMP-impregnated beds, the CO_2 molecules would be attached to the impregnated beds by chemical bonds. Increasing the bed operating temperature would eventually sever these bonds and release CO_2 molecules decreasing their breakthrough time.

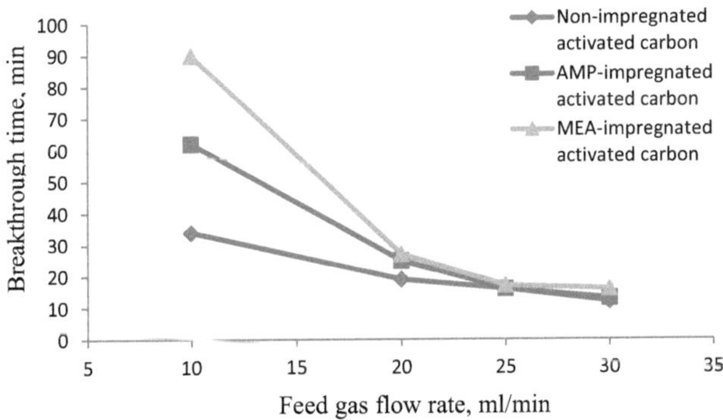

Figure 10 Effect of feed gas flow rate on breakthrough time

3.11. Effect of feed gas flow rate on bed's breakthrough time

Results in Figure 11 are showing the effect of feed gas flow rate at room temperature on the breakthrough time of MEA, AMP-impregnated and non-impregnated AC beds. At high flow rate the reduced residence time would be not enough for chemical reaction to take place between CO_2 and MEA, AMP-impregnated beds reducing their breakthrough time. Increasing feed gas flow rate of non-impregnated AC bed would has minimal effect on their breakthrough time, as there is no chemical reaction and AC particles trapping CO_2 molecules physically. Comparing the results in Figure 10 and 11, it can be concluded that increasing feed gas flow rate had more inverse effect on breakthrough time than increasing operating temperature.

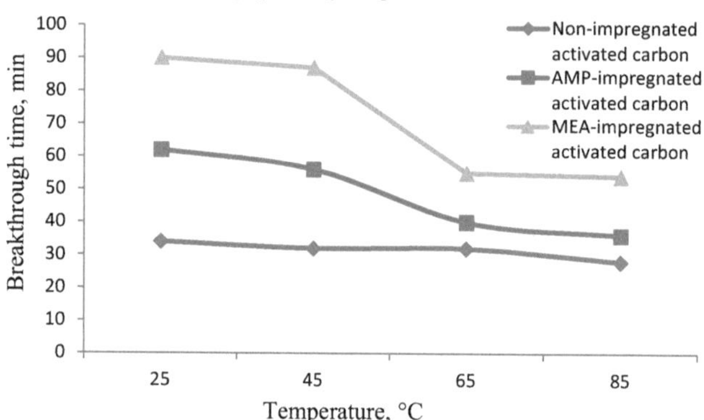

Figure 11 Effect of operating temperature on breakthrough time

4. Conclusion

Adsorption of CO_2 by AC can be enhanced considerably by impregnation with MEA and AMP. Amine impregnation decreased the total micropore volume considerably but at the same time created enormous active sites to adsorb CO_2 molecules. Dynamic breakthrough time of impregnated beds improved significantly comparing to non-impregnated beds. An improvement of 44 % in adsorption bed capacity for 500 μm, AC particles impregnated with MEA comparing to same size AC particles impregnated with AMP. Stoichiometric results are showing that solid amines and in the absence of water are reacting in contrary mode to what is a well-known behavior of liquid amines reaction with CO_2.

Impregnated AC samples became less heterogeneous than non-impregnated AC samples because the micropores in these samples had been blocked by amine based chemicals molecules leading to reducing their pores diversity. Heating up to 75 °C was needed to regenerate exhausted MEA-impregnated beds, while sweeping with N_2 gas flowing at a rate of 60 ml/min for 4 h. at room temperature was enough to regenerate exhausted AMP-impregnated beds.

Because of the exothermic nature of MEA and AMP reaction with CO_2, increasing surrounding temperature would have unfavorable effect on CO_2 adsorption for impregnated beds and less on non-impregnated beds. Increasing CO_2 gas flow rate had reduced breakthrough time more than increasing beds temperature, particularly for impregnated beds. Unlike exhausted AMP-impregnated beds, heat required to regenerate exhaused MEA-impregnate beds.

Acknowledgment

The authors would like to thank University of Malaya for offering the necessary fund for this research through the, University of Malaya Research Grant UMRG RP15/2012A.

References

[1] Vijayavenkataraman S., Iniyan, S., Goic R., 2012. A review of climate change, mitigation and adaptation. Renewable and sustainable energy reviews. 16, 878–897.

[2] Bodri, L. Cermak, V., 2007. Borehole Climatology: A new method how to reconstruct climate. 1st ed. Elsevier press: Amsterdam.

[3] Florides G.A., Christodoulides P., 2009. Global warming and carbon dioxide through sciences. Environment International. 35, 390 401.

[4] Srirangan K., Akawi L., Moo-Young M., Chou C. P., 2012. Towards sustainable production of clean energy carriers from biomass resources. Applied Energy. 100, 172–86.

[5] Chiari L., Zecca A., 2011. Constraints of fossil fuels depletion on global warming projections. Energy Policy. 39, 5026–54.

[6] Lua A.C., Jia Q., 2009. Adsorption of phenol by oil–palm-shell activated carbons in a fixed bed. Chem. Eng. J. 150, 455–61.

[7] Wang M., Lawal A., Stephenson P., Sidders J., Ramshaw C., 2011. Post-combustion CO2 capture with chemical absorption: A state-of-the-art review. Chem. Eng. Research and Design. 89,

1609–24.Edited By Adrian C Todd, Bahman Tohidi, Arne Fredheim, Geoffrey Maitland, Paul Fennell, Stefano Brandani, Jon Gibbins and Geoffrey StevensTop of FormBottom of FormEdited By Aimar Pierre and Nunes Suzana

[8] Brunetti A., Scura F., Barbieri G., Drioli E., 2010. Membrane technologies for CO_2 separation. Journal of Membrane Science. 359, 115–25.

[9] Tuinier M.J., Hamers H.P., van Sint Annaland M., 2011. Techno-economic evaluation of cryogenic CO_2 capture-A comparison with absorption and membrane technology. Int. J. Greenhouse Gas Control. 5: 1559–65.

[10] Duffy A., Walker G.M., Allen S.J., 2006. Investigation on the adsorption of acidic gases using activated dolomite. Chem. Eng. J. 117, 239–44.

[11] Alfe M., Ammendola P., Gargiulo V., Raganati F., Chirone R., 2014. Magnetite loaded carbon fine particles as low-cost CO_2 adsorbent in a sound assisted fluidized bed. Proceedings of the Combustion Institute. DOI: 10.1016/j.proci.2014.06.037.

[12] Zhang F., Li G-D., Chen J-S. 2008. Effects of raw material texture and activation manner on surface area of porous carbons derived from biomass resources. J. of Colloid and Interface Sci. 327, 108–14.

[13] Figueiredo L.J., Pereira M.F.R., Freitas M.M.A., Orfaro J. J. M.,. 1999. Modification of the surface chemistry of activated carbons. Carbon. 37, 1379-89.

[14] Rodriguez-Reinoso F., Molina-Sabio M., 1998. Textural and chemical characterization of microporous carbons. Adv. in Colloid and Interface Sci. 67-77, 271-94.

[15] Li K., Ling L., Lu C., Qiao W., Liu Z., Liu L., Mochida I,. 2001. Catalytic removal of SO_2 over ammonia-activated carbon fibers. Carbon. 39, 1803-08.

[16] Caglayan B.S., Aksoylu A.E., 2013. CO_2 adsorption on chemically modified activated carbon. Journal of Hazardous Materials. 252–253, 19–28.

[17] Plaza M.G., Pevida C., Arenillas A., Rubiera F., Pis J.J., 2007. CO_2 capture by adsorption with nitrogen enriched carbons. Fuel. 86, 2204–12.

[18] Song H.K., Lee K.H., 1998. Adsorption of carbon dioxide on chemically modified carbon adsorbents. Sep. Sci. Technol. 33, 2039–57.

[19] Aroua M.K., Daud W.M.A.W., Yin C.Y., Donni A., 2008. Adsorption capacities of carbon dioxide, oxygen, nitrogen and methane on carbon molecular basket derived from polyethyleneimine impregnation on microporous palm shell activated carbon. Separation and Purification Technology. 62, 609–13.

[20] Zhang M., Guo Y., 2013. Rate based modeling of absorption and regeneration for CO2 capture by aqueous ammonia solution. Applied Energy. 111, 142-52.

[21] Gaspar J., Cormos A-M., 2012. Dynamic modeling and absorption capacity assessment of CO2 capture process. International Journal of Greenhouse Gas Control. 8, 45–55.

[22] Rubin E.S., Mantripragada H., Marks A., Versteeg P., Kitchin J., 2012. The outlook for improved carbon capture technology, Review. Progress in Energy and Combustion Science. 38, 630–71.

[23] Raganati F., Ammendola P., Chirone R., 2014. CO2 adsorption on fine activated carbon in a sound assisted fluidized bed: Effect of sound intensity and frequency, CO2 partial pressure and fluidization velocity. Applied Energy. 113, 1269–82.

[24] Guo J., Lua A.C., 2003. Textural and chemical properties of adsorbent prepared from palm shell by phosphoric acid activation. Materials Chemistry and Physics. 80, 114–19.

[25] Strube R., Pellegrini G., Manfrida G., 2011. The environmental impact of post-combustion CO2 capture with MEA, with aqueous ammonia, and with an aqueous ammonia-ethanol mixture for a coal-fired power plant. Energy. 36, 3763–70.

[26] Rodriguez-Flores H.A., Mello L.C., Salvagnini W.M., de Paiva J.L., 2013. Absorption of CO2 into aqueous solutions of MEA and AMP in a wetted wall column with film promoter. Chemical Engineering and Processing: Process Intensification. 73, 1–6.

[27] Yu C.H., Huang C.H., Tan C.S., 2012. A review of CO2 capture by absorption and adsorption. Aerosol and Air Quality Research. 12, 745-69.

[28] Khalil S.H., Aroua M.K., Daud W.M.A.W., 2012. Study on the improvement of the capacity of amine-impregnated commercial activated carbon beds for CO2 adsorbing. Chem. Eng. J. 183, 15–20.

[29] Deepatana A., Valix M., 2006. Recovery of nickel and cobalt from organic acid complexes: Adsorption mechanisms of metal-organic complexes onto aminophosphonate chelating resin. J. Hazard. Mater. 37, 925–33.

[30] Zhao Y., Ho W.S.W., 2012. Steric hindrance effect on amine demonstrated in solid polymer membranes for CO2 transport. J. of Membrane Sci. 415–416, 132–38.

[31] Chakraborty A.K., Astarita G., Bischoff K.B., 1986. CO2 absorption in aqueous solutions of hindered amines, Chem. Eng. Sci. 41, 997–1003.

[32] Tontiwachwuthikul P., Meisen A., Lim C.J., 1992. CO2 absorption by NaOH, mono ethanolamine and 2-amino-2-methyl-1-propanol solutions in a packed column, Chem. Eng. Sci. 47, 381–90.

[33] Park J.Y., Yoon S.J., Lee H., 2003. Effect of steric hindrance on carbon dioxide absorption into new amine solutions: thermodynamic and spectroscopic verification through solubility and NMR analysis. Environ. Sci. Technol. 37, 1670–75.

[34] Sahaa A.K., Biswasb A.K., Bandyopadhyayc S.S., 1999 Absorption of CO2 in a sterically hindered amine: modeling absorption in a mechanically agitated contactor. Separation and purification tech. 15, 101–12

[35] Sing K.S.W., Everett D.H., Haul R.A.W., Moscou L., Pierotti R.A., Rouquerol J.,

Siemieniewska T., 1985. Reporting physisorption data for gas/solid systems with special reference to the determination of surface area and porosity. Pure Appl. Chem. 57, 606-12.

[36] Everett D.H., Powl J.C., 1976. Adsorption in slit-like and cylindrical micropores in the henry's law region. A model for the microporosity of carbons. J. Chem. Soc. Faraday Trans. 1: Physical Chemistry in Condensed Phases. 72, 619-36.

[37] Amankwah K.A.G., Schwakz J.A., 1995. A modified approach for estimating pseudo-vapor pressures in the application of the Dubinin-Astakhov equation. Carbon. 33, 1313-19.

[38] Puziy A.M., 1994. Heterogeneity of Synthetic Active Carbons. Langmuir. 11, 543-44.

[39] Garnier C., Gorner T., Villieras F., De Donato P.h., Polakovic M., Bersillon J-L., 2007. Activated carbon surface heterogeneity seen by parallel probing by inverse liquid chromatography at the solid/liquid interface and by gas adsorption analysis at the solid/gas interface . Carbon. 45, 240–47.

[40] Wu Y., Carroll J.J., Du Z., 2011. Editors. Carbon dioxide sequestration and related technologies, Salem, Massachusetts: LLC Publishing. Scrivener.

[41] Whitten K.W., Davis R.E., Peck M.L., Stanley G.G., 2009. Chemistry.9th ed. California: Brooks/cole. Cencage learning.

YOUR KNOWLEDGE HAS VALUE

- We will publish your bachelor's and
  master's thesis, essays and papers

- Your own eBook and book -
  sold worldwide in all relevant shops

- Earn money with each sale

Upload your text at www.GRIN.com
and publish for free